Make it Easy

CW00411577

Age 6-7

Maths

Contents

Learning Activities

2 Numbers to 20
3 Counting
4 Adding
5 2-D shapes
6 Taking away
7 Numbers to 100
8 Addition and subtraction
9 Counting patterns
10 Measuring length
11 Finding totals
12 Odd and even numbers
13 3-D shapes
14 Measuring mass
15 Breaking up numbers
16 Reading the time

17 Multiplying
18 Symmetrical shapes
19 Ordering numbers
20 Dividing
21 Reading graphs
22 Time
23 2 times table
24 Half fractions
25 Measuring capacity
26 Money
27 Number sequences
28 Multiplication facts
29 Quarter fractions
30 Problems
31 Finding the difference

Quick Tests

32 Test 1 Read and write numbers to 100
33 Test 2 Addition
34 Test 3 Money: totalling
35 Test 4 2-D shapes
36 Test 5 Counting sequences within 50
37 Test 6 Subtraction: finding differences
38 Test 7 Multiplication: repeated addition
39 Test 8 Division: sharing
40 Test 9 Time (1)
41 Test 10 Data: block graphs
42 Test 11 Breaking up numbers
43 Test 12 Subtraction facts
44 Test 13 Money: giving change
45 Test 14 3-D shapes

46 Test 15 Counting patterns
47 Test 16 Decade sums
48 Test 17 2 times table
49 Test 18 Fractions: halves and quarters
50 Test 19 Measures: length
51 Test 20 Data: pictograms
52 Test 21 Comparing and ordering numbers
53 Test 22 Addition and subtraction
54 Test 23 Problems
55 Test 24 Shapes
56 Test 25 Odd and even numbers
57 Test 26 Money and place value
58 Test 27 10 times table
59 Test 28 Fractions of quantities
60 Test 29 Time (2)
61 Test 30 Data: tables

Answers

62 Answers

Paul Broadbent and Peter Patilla

Numbers to 20

The numbers between 12 and 20 are **teen** numbers.

They all end in **...teen**.

11 and 12 are made from a ten and ones,
but do not end in **teen**.

13 ➜ thirteen
10 + 3 = 13

I Write the words or numbers for each of these.

a 15 ➜

b 18 ➜

c 11 ➜

d 17 ➜

e fourteen ➜

f nineteen ➜

g twelve ➜

h sixteen ➜

II Write the word for each number.

a 13 ➜

b 12 ➜

c 18 ➜

d 17 ➜

e 14 ➜

f 19 ➜

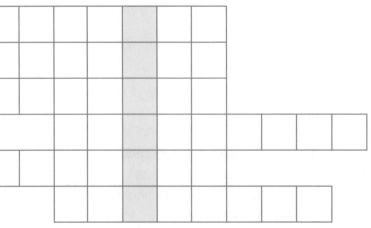

The hidden number in the shaded area is ☐.

Counting

Use this grid to help you learn the **order** of numbers to 50.

1	2	3	4	5	6	7	8	9	10
11	12	13	14	15	16	17	18	19	20
21	22	23	24	25	26	27	28	29	30
31	32	33	34	35	36	37	38	39	40
41	42	43	44	45	46	47	48	49	50

I Fill in the missing numbers.

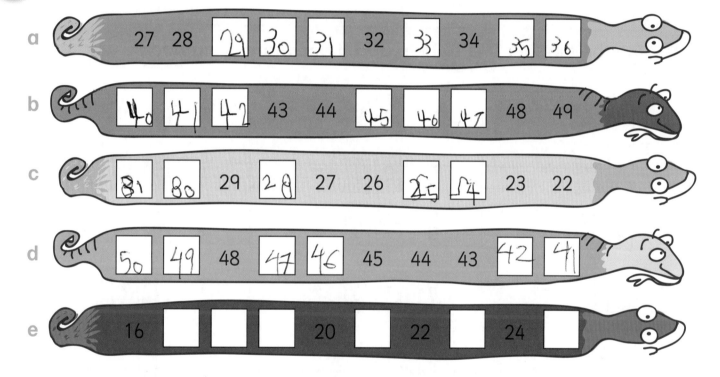

a 27 28 **29 30 31** 32 **33** 34 **35 36**

b **40 41 42** 43 44 **45 46 47** 48 49

c **31 80** 29 **28** 27 26 **25 24** 23 22

d **50 49** 48 **47 46** 45 44 43 **42 41**

e 16 ☐ ☐ ☐ 20 ☐ 22 ☐ 24 ☐

II These are all part of the 1–50 grid. Use the grid at the top of the page to help fill in the missing numbers.

a

14			
	26		28
	35		

b

22			
32	33		36
		44	

c

			10
		18	
		38	40

Adding

A **number line** can help with addition.

What is 4 + 7?

Start with the biggest number and count on.

7 + 4 = 11

I **Use the number line to help add these pairs of numbers.**

a 6 3 → ☐

b 5 7 → ☐

c 8 4 → ☐

d 7 2 → ☐

e 6 6 → ☐

f 9 5 → ☐

g 4 9 → ☐

h 7 8 → ☐

i 3 8 → ☐

j 6 7 → ☐

k 8 5 → ☐

l 7 7 → ☐

II **Draw a line from each addition problem to its total. Colour the star with no matching fact.**

a 3 + 9

c 7 + 7

b 9 + 9

d 9 + 6

e 8 + 8

f 7 + 6

g 7 + 3

h 10 + 10

i 10 + 9

j 6 + 5

10 11 12 13 14 15 16 17 18 19 20

2-D shapes

A 2-D shape is a **flat shape**.

Learn the names and number of sides of these shapes.

| triangle | quadrilateral | pentagon | hexagon | heptagon | octagon |
| 3 sides | 4 sides | 5 sides | 6 sides | 7 sides | 8 sides |

Rectangles and squares are special quadrilaterals.

I **Draw lines to join the shapes to the correct name.**

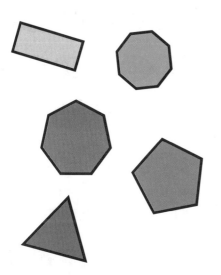

triangle

rectangle

pentagon

hexagon

heptagon

octagon

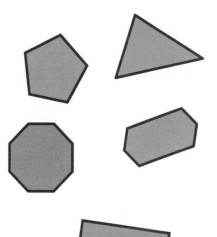

II **Colour this stained glass window using the colour code below.**

 triangles

 quadrilaterals

 pentagons

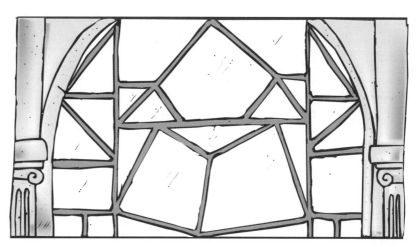

Taking away

You can count back along a number line to help you **subtract**, or **take away**.

What is 14 – 5? Start at 14 and count back 5.

14 – 5 = 9

I Show the jumps for each subtraction. Then write the answer in the box.

a 11 – 4 = ☐   ①②③④⑤⑥⑦⑧⑨⑩⑪⑫⑬⑭⑮

b 13 – 5 = ☐   ①②③④⑤⑥⑦⑧⑨⑩⑪⑫⑬⑭⑮

c 12 – 7 = ☐   ①②③④⑤⑥⑦⑧⑨⑩⑪⑫⑬⑭⑮

d 16 – 7 = ☐   ⑥⑦⑧⑨⑩⑪⑫⑬⑭⑮⑯⑰⑱⑲⑳

e 18 – 6 = ☐   ⑥⑦⑧⑨⑩⑪⑫⑬⑭⑮⑯⑰⑱⑲⑳

f 17 – 6 = ☐   ⑥⑦⑧⑨⑩⑪⑫⑬⑭⑮⑯⑰⑱⑲⑳

II Make each subtraction match the number on the stars.

a

10 – ☐

☐ – 2

12 – ☐

☐ – 3

☐ – 8

b

11 – ☐

☐ – 7

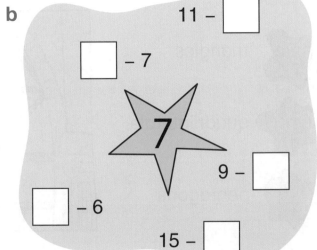

9 – ☐

☐ – 6

15 – ☐

Numbers to *100*

This grid shows the numbers to 100.

Use the **tens** to help you read and write the numbers.

20 twenty	60 sixty
30 thirty	70 seventy
40 forty	80 eighty
50 fifty	90 ninety

1	2	3	4	5	6	7	8	9	10
11	12	13	14	15	16	17	18	19	20
21	22	23	24	25	26	27	28	29	30
31	32	33	34	35	36	37	38	39	40
41	42	43	44	45	46	47	48	49	50
51	52	53	54	55	56	57	58	59	60
61	62	63	64	65	66	67	68	69	70
71	72	73	74	75	76	77	78	79	80
81	82	83	84	85	86	87	88	89	90
91	92	93	94	95	96	97	98	99	100

I Circle the correct number for each of these.

a thirty-eight   83   37   78   38   30

b fifty-four   44   50   46   45   54

c seventy-nine   97   17   79   76   96

d sixty-two   52   26   60   62   80

e eighty-seven   81   78   80   76   87

II Find these numbers on the word search. They are written across → and down ↓.

20	60
30	70
40	80
50	90

T	W	E	N	T	Y	E	F
N	S	I	X	T	Y	F	I
I	Y	G	N	H	V	O	F
N	E	H	E	Y	I	R	T
E	S	T	H	I	R	T	Y
T	R	Y	M	L	F	Y	E
Y	S	E	V	E	N	T	Y

Addition and subtraction

This **number trio** can make 4 addition and subtraction facts.

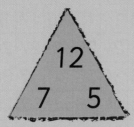

7 + 5 = 12   12 – 7 = 5

5 + 7 = 12   12 – 5 = 7

 Fill in the addition and subtraction facts for these.

a

5	+		=	
	+	5	=	
	–	5	=	
	–		=	5

b

	+		=	15
	+	6	=	15
15	–		=	
15	–		=	

c

9	+		=	
	+	9	=	
	–		=	9
	–	9	=	

 Choose 8 different numbers from the line below to complete these facts.

 4 + ☐ = 9

 7 – 5 = ☐

 8 – ☐ = 1

 ☐ + ☐ = 11

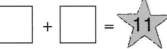 12 – ☐ = 8

 ☐ + 3 = ☐

8

Counting patterns

Practise **counting on** and **back** in steps of 2, 5 and 10.

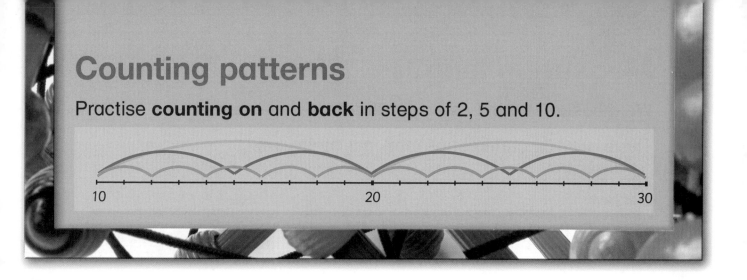

10   20   30

I Continue each of these counting patterns to 100. Mark them on the 100 square like this:

2 4 6 8 10 12 → 100

5 10 15 20 25 → 100

10 20 30 40 50 → 100

1	2	3	4	5	6	7	8	9	10
11	12	13	14	15	16	17	18	19	20
21	22	23	24	25	26	27	28	29	30
31	32	33	34	35	36	37	38	39	40
41	42	43	44	45	46	47	48	49	50
51	52	53	54	55	56	57	58	59	60
61	62	63	64	65	66	67	68	69	70
71	72	73	74	75	76	77	78	79	80
81	82	83	84	85	86	87	88	89	90
91	92	93	94	95	96	97	98	99	100

II Count in 5s and fill in the next 4 numbers.

a   4   9

b   22 27

c   43   48

Now count in 10s and fill in the next 4 numbers.

d   8   18

e   27 37

f 39 49

9

Measuring length

We measure lengths using **centimetres** and **metres**.

There are 100 centimetres (cm) in 1 metre (m).

100 cm = 1 m

```
0  1  2  3  4  5  6  7  8  9  10  11  12  13  14  15
```

This line shows 15 cm.

I   **Use a ruler to measure each of these lengths in centimetres.**

a   ☐ cm

b   ☐ cm

c   ☐ cm

d   ☐ cm

e   ☐ cm

f   ☐ cm

> Try estimating the length before you measure.

II   **Draw lines to join these objects to the most likely length.**

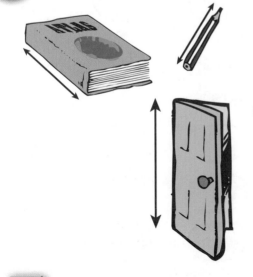

about 1 metre

about 2 metres

about 10 centimetres

more than 2 metres

about 50 cm

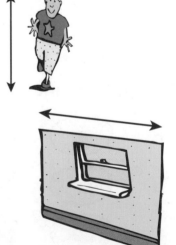

Finding totals

When you add together 3 or more numbers, try **starting** with the **largest number**.

Here's how to total 4, 3 and 8:

$$8 + 4 = 12$$

$$12 + 3 = 15$$

You could look for pairs that are easy to total.

Here's how to total 6, 5 and 4:

$$6 + 4 = 10$$

$$10 + 5 = 15$$

I Fill in the totals for these sets of additions.

a (6 2 8) → ☐   d (9 1 8) → ☐   g (8 4 6) → ☐

b (5 9 2) → ☐   e (7 4 3) → ☐   h (3 3 8) → ☐

c (7 2 3) → ☐   f (5 9 5) → ☐   i (6 4 3) → ☐

II Make these totals in different ways.

a $4 + \square + \square$

$\square + \square + 5$

$\square + 6 + \square$

13

$\square + 3 + \square$

$8 + \square + \square$

$\square + \square + 1$

b $\square + \square + 9$

$\square + 6 + \square$

$7 + \square + \square$

18

$4 + \square + \square$

$\square + 8 + \square$

$\square + \square + 5$

Odd and even numbers

Even numbers always end in

2 4 6 8 0

36

is an even number.

Odd numbers always end in

1 3 5 7 9

63

is an odd number.

I **Write the next even number.**

a 22 →

b 38 →

c 46 →

d 60 →

e 54 →

Write the next odd number.

f 35 →

g 57 →

h 29 →

i 87 →

j 91 →

II **Colour red the even number trail. Start at the IN gate.**

OUT

 IN

19	24	32	48	85	33	34	26	18	70	96	73	34	26	14
23	6	61	16	51	27	58	35	43	19	34	85	58	21	43
42	30	25	40	10	7	94	65	24	46	52	17	92	80	19
85	27	41	93	28	43	62	97	12	21	33	29	31	52	21
17	35	43	8	32	76	44	81	16	54	36	28	56	74	45

How many stars have you collected?

3-D shapes

A 3-D shape is a **solid shape**.

Learn the names of these shapes. Look at the faces.

square face

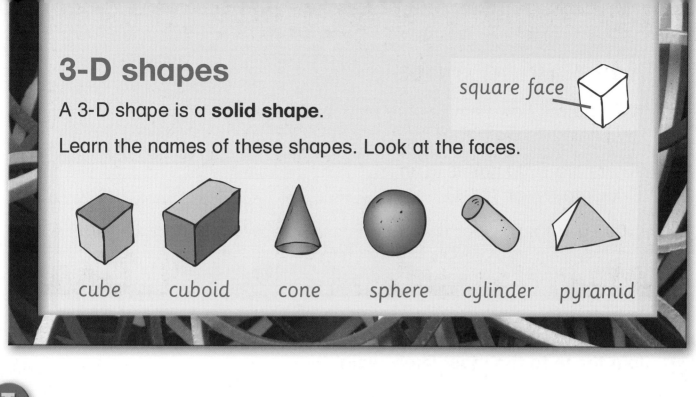

cube   cuboid   cone   sphere   cylinder   pyramid

I Draw lines to join each shape to its correct name.

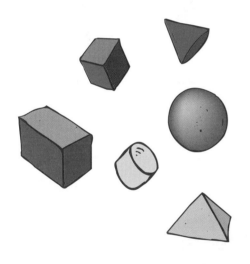

sphere

cylinder

cone

cuboid

pyramid

cube

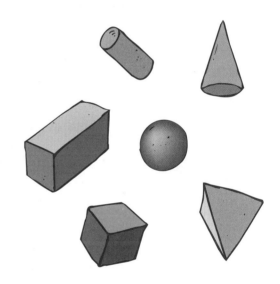

II Fill in the number of faces for each of these shapes.

a

☐ square faces

☐ rectangle faces

c

☐ square faces

☐ triangle faces

b

☐ square faces

d

☐ circle faces

and a curved face

Measuring mass

We find out how heavy something is by finding its **weight** or **mass**.

There are 1000 grams (g) in 1 kilogram (kg).

1000 g = 1 kg

Find something that weighs about 1 kg.

I Join these to the most likely weight.

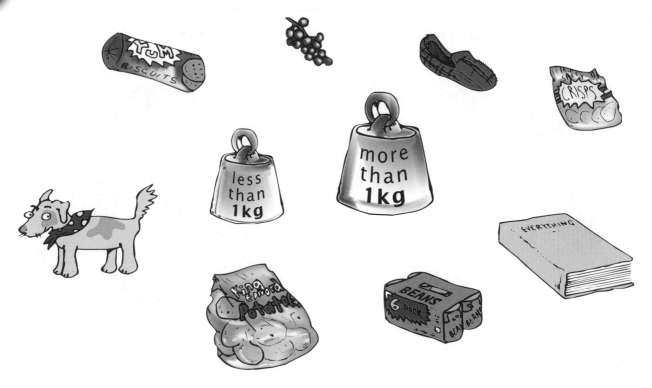

II Write down the weight on the scales to the nearest kilogram.

a   b   c   d

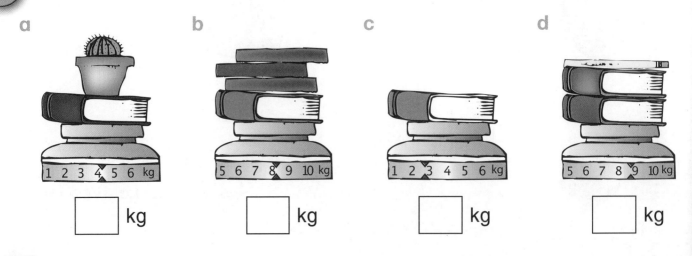

☐ kg   ☐ kg   ☐ kg   ☐ kg

Breaking up numbers

The numbers between **10** and **99** all have **two digits**.

$$57 \rightarrow 50 + 7$$

5 tens   7 ones

I   **Fill in the missing numbers.**

a 34 $\rightarrow$ 30 + ☐   d 65 $\rightarrow$ ☐ + 5   g 29 $\rightarrow$ 20 + ☐

b 51 $\rightarrow$ ☐ + 1   e 83 $\rightarrow$ 80 + ☐   h 76 $\rightarrow$ ☐ + 6

c 47 $\rightarrow$ 40 + ☐   f 42 $\rightarrow$ ☐ + 2   i 59 $\rightarrow$ 50 + ☐

II   **Draw lines to join the matching pairs.**

Reading the time

These clocks show **quarter past eight.**

8:15

15 minutes have gone past 8 o'clock.

These clocks show **quarter to five.**

4:45

45 minutes have gone past 4 o'clock.

I Draw lines to join the clocks showing the same time.

a   b   c   d   e   f   g

| 11:30 | 3:00 | 1:15 | 12:45 | 6:45 | 4:15 | 7:15 |

II Write the number of minutes between each of these times.

a  [ ] minutes

c 10:45 11:00 [ ] minutes

b  [ ] minutes

d 3:15 3:45 [ ] minutes

Multiplying

Counting in equal groups is also called **multiplying**.

The multiplication sign is **×**.

$3 + 3 + 3 + 3 = 12$

4 lots of 3 is 12

$4 \times 3 = 12$

$4 + 4 + 4 = 12$

3 lots of 4 is 12

$3 \times 4 = 12$

These both give the same answer.

I **Write the answers to these facts.**

a

$2 + 2 + 2 = \boxed{}$

$3 \times 2 = \boxed{}$

d

$4 + 4 = \boxed{}$

$2 \times 4 = \boxed{}$

b

$5 + 5 = \boxed{}$

$2 \times 5 = \boxed{}$

e

$3 + 3 + 3 = \boxed{}$

$3 \times 3 = \boxed{}$

c

$3 + 3 + 3 + 3 + 3 = \boxed{}$

$5 \times 3 = \boxed{}$

f

$4 + 4 + 4 + 4 = \boxed{}$

$4 \times 4 = \boxed{}$

II **Draw 2 spots on each hat. Then write the answer.**

Draw 3 spots on each hat. Then write the answer.

a

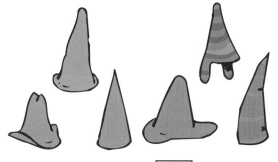

$6 \times 2 = \boxed{}$

b

$6 \times 3 = \boxed{}$

Symmetrical shapes

Shapes are **symmetrical** if they are the same either side of a **mirror line**.

The mirror line is called the **line of symmetry**.

I Draw a line of symmetry on each of these shapes.

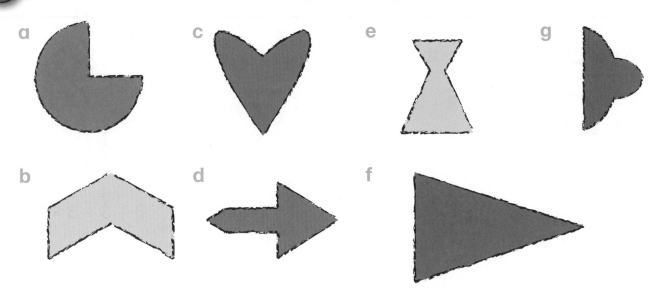

a

c

e

g

b

d

f

II Complete this to make a symmetrical shape. Then colour it to make a symmetrical pattern.

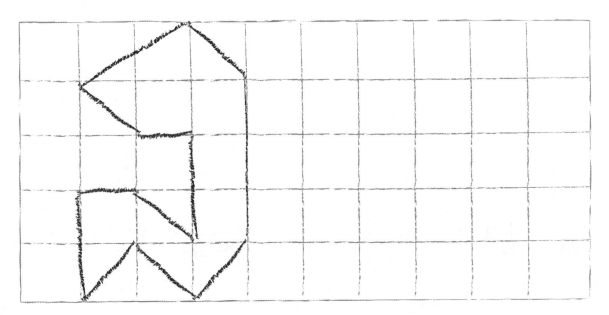

Ordering numbers

When you put **2-digit** numbers in order, look at the **tens and then the ones** digit.

52 is larger than 38 because 5 tens is more than 3 tens.

Use this number line to help.

I Write a number in each box so there are four numbers in order.

a 42 ⟩ ⟩ ⟩ 50 ⟩   d 82 ⟩ ⟩ ⟩ 91 ⟩   g 94 ⟩ ⟩ ⟩ 98 ⟩

b 61 ⟩ ⟩ ⟩ 73 ⟩   e 66 ⟩ ⟩ ⟩ 71 ⟩   h 88 ⟩ ⟩ ⟩ 92 ⟩

c 57 ⟩ ⟩ ⟩ 64 ⟩   f 76 ⟩ ⟩ ⟩ 83 ⟩   i 57 ⟩ ⟩ ⟩ 61 ⟩

II Write these sets in order, starting with the smallest amount.

a

58p 39p 85p 61p 42p

□ p □ p □ p □ p □ p

b
73 kg 69 kg 37 kg 39 kg 76 kg

□ kg □ kg □ kg □ kg □ kg

c
32 cm 23 cm 80 cm 38 cm 28 cm

□ cm □ cm □ cm □ cm □ cm

d
£53 £62 £29 £65 £35

£□ £□ £□ £□ £□

19

Dividing

Dividing a number of objects can be shown by grouping them.

The division sign is ÷.

6 beans grouped in twos, gives 3 groups

$$6 \div 2 = 3$$

 Draw loops around these beans to group them. Write the answers.

a 8 grouped in 2s

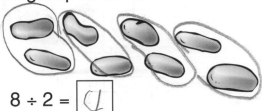

8 ÷ 2 = 4

c 12 grouped in 3s

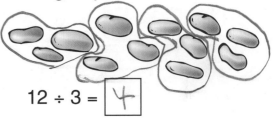

12 ÷ 3 = 4

b 9 grouped in 3s

9 ÷ 3 = 3

d 10 grouped in 2s

10 ÷ 2 = 5

 Use the cookies to help you complete these.

a

15 ÷ 5 = 3

c

10 ÷ 2 = 5

b

15 ÷ 3 = 5

d

10 ÷ 5 = 2

20

Reading graphs

Block graphs show information in a simple way.

Count the blocks carefully or read across for the amount.

A group of children threw 10 beanbags trying to get them into a bucket.

How many more beanbags did Zoe get in than Fred?

I A group of children tested how many pegs they could hold in one hand.

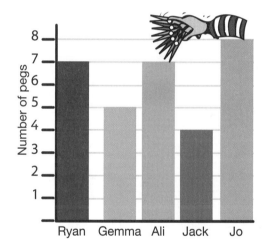

a Who held the most pegs?

b How many pegs did Gemma hold?

c Which 2 children held the same number of pegs?

d Who held the fewest pegs?

e How many more pegs did Jo hold than Jack?

II Carry out your own peg test. Ask family and friends to hold as many pegs as they can in one hand. Record your results as a pictogram.

Name	Number of pegs

= 1 peg

Time

There are **12 months** or **52 weeks** in **a year**.

Try to learn the order of the months and the seasons. Think about the month and season you were born in.

 spring summer autumn winter

I Complete the names of the months for each season.

spring

M __ __ c h
A __ r __ l
__ __ y

winter

D __ __ __ m b __ __
__ __ n u __ r __
F __ __ r __ __ __ y

summer

J __ __ e
__ __ __ y
A __ __ u __ __

autumn

S __ p __ __ __ __ __ __ r
__ __ t __ b __ __
N __ __ e __ b __ __

II Complete these time facts.

a ☐ days in a week

b ☐ months in a year

c ☐ weeks in a year

d ☐ hours in a day

e ☐ minutes in an hour

f ☐ seconds in a minute

g ☐ days in a fortnight

h ☐ days in a weekend

i ☐ seasons in a year

j ☐ months in a season

2 times table

The numbers in the **2 times table** can be shown as a pattern.

Try to learn the 2 times table by heart.

$1 \times 2 = 2$

$2 \times 2 = 4$

I Draw lines to join the questions to the correct answers.

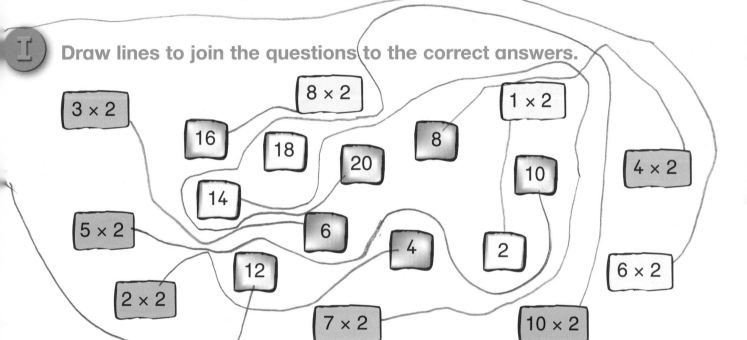

8 × 2   3 × 2   16   18   20   14   5 × 2   6   4   12   2 × 2   7 × 2   1 × 2   8   10   2   4 × 2   6 × 2   10 × 2

a Write a calculation for the answer that is left. 9×2

$9 \times 2 = 18$

II Answer these questions as fast as you can. Ask someone to time you.

a 3 × 2 = ☐ 6

7 × 2 = ☐ 8 14

4 × 2 = ☐ 8

1 × 2 = ☐ 2

6 × 2 = ☐ 12

b 2 × 10 = ☐ 20

2 × 2 = ☐ 4

2 × 8 = ☐ 16

2 × 5 = ☐ 10

2 × 9 = ☐ 18

c 8 × 2 = ☐ 12

2 × 6 = ☐ 12

9 × 2 = ☐ 18

2 × 7 = ☐ 14

5 × 2 = ☐ 10

2 mins 32 secs.

23

Half fractions

This chocolate bar is cut into **2 equal pieces**.

Each piece is **half ($\frac{1}{2}$)** of the whole bar.

There are 8 squares of chocolate.

$\frac{1}{2}$ of 8 = 4

 Colour $\frac{1}{2}$ of each shape.

a

c

e
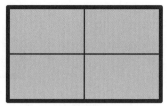

b
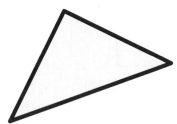

d

f
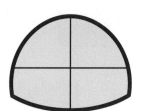

 Circle $\frac{1}{2}$ of each set. Write the answer.

a

$\frac{1}{2}$ of 6 = ☐

c

$\frac{1}{2}$ of 10 = ☐

b
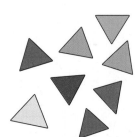
$\frac{1}{2}$ of 8 = ☐

d

$\frac{1}{2}$ of 4 = ☐

Measuring capacity

The **capacity** of a jug shows how much **liquid** it holds.

> 1000 millilitres (ml) = 1 litre (l)

Fill a 1 litre jug so that you know how much a litre is.

I Draw lines to join these things to the most likely amount.

less than 1 litre	greater than 1 litre

II Write down these amounts to the nearest litre.

a

☐ litres

c

☐ litres

e

☐ litres

b

☐ litres

d

☐ litres

f

☐ litres

Money

Practise finding totals of **coins** and giving **change**. When you give change, try counting up.

A cake costs 39p. Think about the change you will get from 50p.

Count on from 39p to 50p.

+1p   +10p

39p   40p   50p

The change is 11p.

 Draw the 3 coins you would use to buy each of these.

a ◯ ◯ ◯   d  ◯ ◯ ◯

b  ◯ ◯ ◯   e  ◯ ◯ ◯

c  ◯ ◯ ◯   f  ◯ ◯ ◯

 This is the change given from 50p. How much did each cake cost?

a

cost ➜ ☐ p

b

cost ➜ ☐ p

c

cost ➜ ☐ p

Number sequences

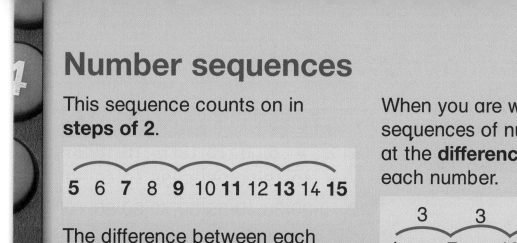

This sequence counts on in **steps of 2**.

5 6 **7** 8 **9** 10 **11** 12 **13** 14 **15**

The difference between each number is 2.

When you are writing sequences of numbers, look at the **difference** between each number.

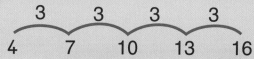

| | 3 | | 3 | | 3 | | 3 | |
| 4 | | 7 | | 10 | | 13 | | 16 |

I Write the next 2 numbers in each sequence.

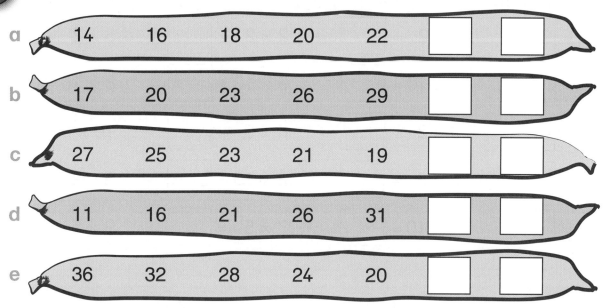

a   14   16   18   20   22   ☐ ☐

b   17   20   23   26   29   ☐ ☐

c   27   25   23   21   19   ☐ ☐

d   11   16   21   26   31   ☐ ☐

e   36   32   28   24   20   ☐ ☐

II Write 3 sequences of your own. The number 20 must be in each sequence.

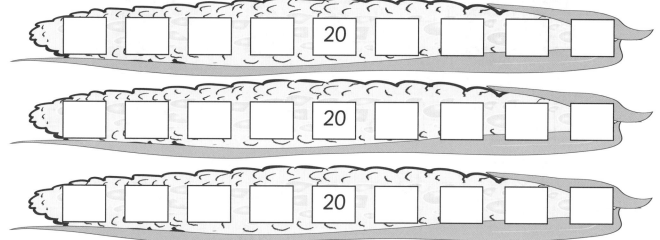

Multiplication facts

Try to learn the **2** times,
5 times and **10** times tables.

Use this grid to help you.

×	1	2	3	4	5	6	7	8	9	10
2	2	4	6	8	10	12	14	16	18	20
5	5	10	15	20	25	30	35	40	45	50
10	10	20	30	40	50	60	70	80	90	100

I Cover the grid above. Now answer these questions as fast as you can. Check your answers, then try to beat your score.

a 3×5 = ☐   b 4×2 = ☐   c 2×2 = ☐   d 5×2 = ☐

6×2 = ☐ 7×10 = ☐ 9×10 = ☐ 8×10 = ☐

4×10 = ☐ 9×2 = ☐ 6×5 = ☐ 10×5 = ☐

8×2 = ☐ 5×5 = ☐ 2×10 = ☐ 3×2 = ☐

4×5 = ☐ 3×10 = ☐ 7×5 = ☐ 9×5 = ☐

II Write the digits 1 to 9 in the boxes to make each sum correct.

$2 \times$ ☐ = ☐   ☐ $\times 5 = 20$ ☐ $0 \times$ ☐ $= 80$

☐ $\times$ ☐ $= 35$ ☐ $\times$ ☐ $= 18$

☐ $\times$ ☐ $= 40$

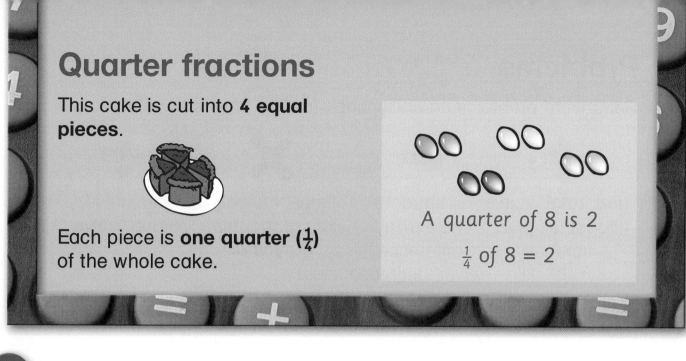

Quarter fractions

This cake is cut into **4 equal pieces**.

Each piece is **one quarter ($\frac{1}{4}$)** of the whole cake.

A quarter of 8 is 2

$\frac{1}{4}$ of 8 = 2

I Colour $\frac{1}{4}$ of each shape.

a
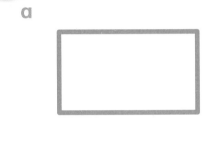

c

e
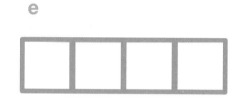

b

d
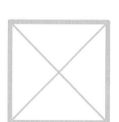

f

II Colour $\frac{1}{4}$ of the ribbons on these badges. Make sure each pattern is different.

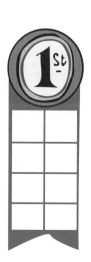

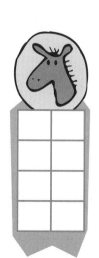

Problems

Read word problems carefully. Look for **key words** to help you.

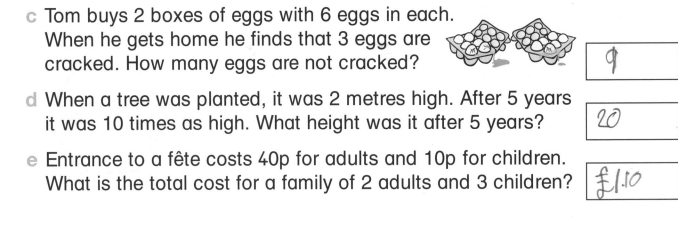

+	−	✗	÷
add, total, sum, altogether, plus, increase	subtract, take away, difference, fewer, decrease	times, multiply, lots of, double, groups of	share, divide, group, halve

I Answer these problems.

a Four friends share 20 pencils equally between them.
How many pencils do they each have?

> 5

b A T-shirt costs £3.50. What change will there be from £5?

> £1.50

c Tom buys 2 boxes of eggs with 6 eggs in each.
When he gets home he finds that 3 eggs are
cracked. How many eggs are not cracked?

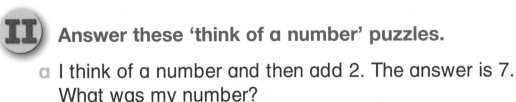

> 9

d When a tree was planted, it was 2 metres high. After 5 years
it was 10 times as high. What height was it after 5 years?

> 20

e Entrance to a fête costs 40p for adults and 10p for children.
What is the total cost for a family of 2 adults and 3 children?

> £1.10

II Answer these 'think of a number' puzzles.

a I think of a number and then add 2. The answer is 7.
What was my number?

> 5

b I think of a number and then take away 5. The answer is 6.
What was my number?

> 11

c I think of a number and then halve it. The answer is 4.
What was my number?

> 8

d I think of a number and then double it. The answer is 10.
What was my number?

> 5

Finding the difference

To find the **difference** between 2 numbers, count on from the smaller number.

What is the difference between 19 and 23?

19 20 21 22 23 24 25   23 – 19 = 4

 Write the difference in price between these pairs of items.

a

£8   £13

Difference: £ ☐

c

£24   £19

Difference: £ ☐

e

£37   £41

Difference: £ ☐

b

£25   £17

Difference: £ ☐

d

£33   £28

Difference: £ ☐

f

£46   £52

Difference: £ ☐

 Draw lines to join pairs with a difference of 6.

62

56

94

88

74

81

81

75

83

77

68

Test 1 Read and write numbers to 100

Use these numbers to help you.

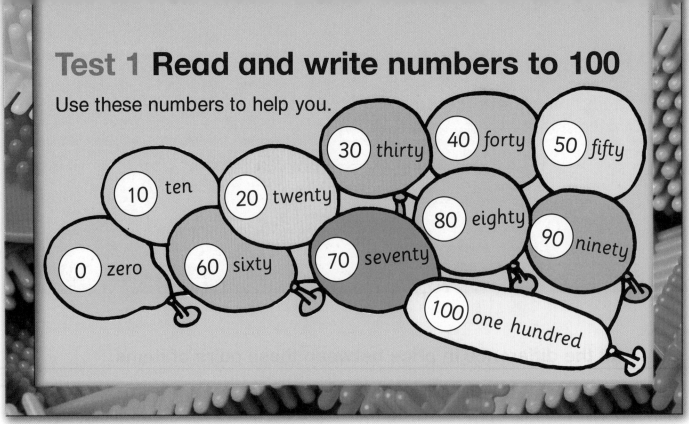

Write the numbers to match the words.

1. thirty-four

2. forty-six

3. twenty-eight

4. seventy-two

5. eighty-nine

Write these numbers as words.

6. 23 _____

7. 56 _____

8. 91 _____

9. 67 _____

10. 49 _____

Colour in your score

Test 2 Addition

We use a **number line** to help us **add on**.

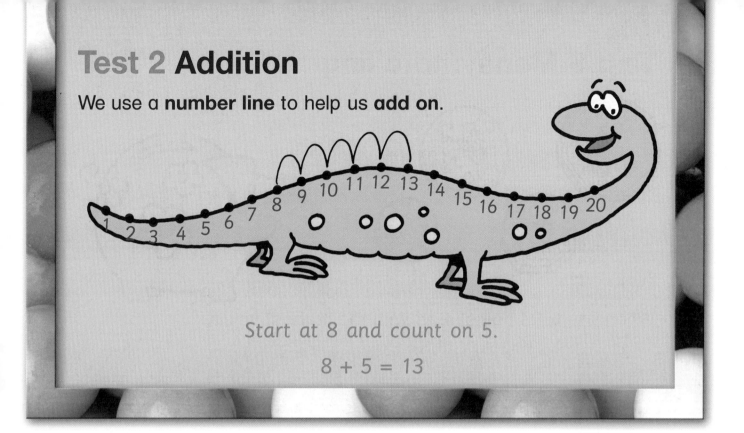

Start at 8 and count on 5.

8 + 5 = 13

Use the number line to help you work out the answers.

1.   6 + 5 =

2. 11 + 3 =

3.   8 + 6 =

4. 12 + 4 =

5.   9 + 5 =

6.   6 + 11 =

7.   7 + 9 =

8.   8 + 4 =

9. 14 + 4 =

10. 13 + 6 =

Colour in your score

33

Test 3 Money: totalling

When you **total coins**, start with the **highest** value.

50p + 20p + 10p + 5p = 85p

Total each set of coins.

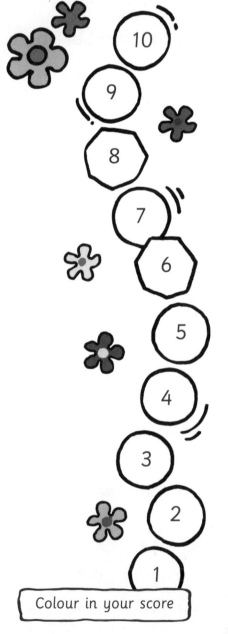

1. (10p) (50p) (2p) ⇨ ☐ p

2. (5p) (2p) (2p) (10p) ⇨ ☐ p

3. (2p) (1p) (20p) (10p) ⇨ ☐ p

4. (20p) (10p) (50p) (2p) ⇨ ☐ p

5. (2p) (5p) (10p) (1p) ⇨ ☐ p

6. (50p) (20p) (2p) (1p) ⇨ ☐ p

7. (20p) (20p) (2p) (10p) (2p) ⇨ ☐ p

8. (2p) (1p) (10p) (5p) (10p) ⇨ ☐ p

9. (50p) (20p) (5p) (10p) (1p) ⇨ ☐ p

10. (2p) (10p) (2p) (5p) (20p) ⇨ ☐ p

Colour in your score

34

Test 4 2-D shapes

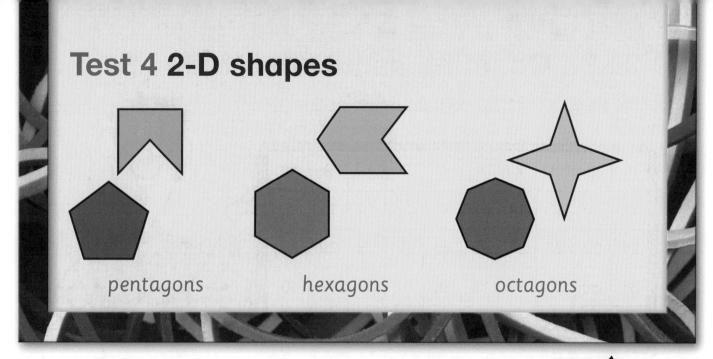

pentagons   hexagons   octagons

Answer these questions.

1. All triangles have ☐ sides.

2. All hexagons have ☐ sides.

3. All pentagons have ☐ sides.

4. All octagons have ☐ sides.

5. All quadrilaterals have ☐ sides.

Name these shapes.

6. _____

7. _____

8. _____

9. _____

10. _____

Colour in your score

35

Test 5 Counting sequences within 50

Use the grid to help with counting sequences.

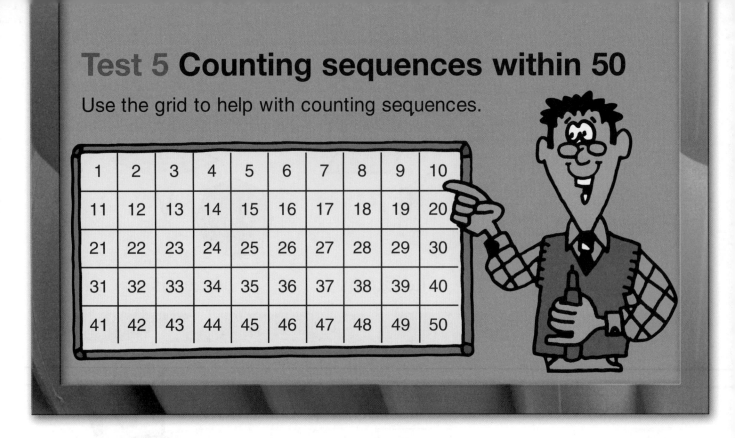

1	2	3	4	5	6	7	8	9	10
11	12	13	14	15	16	17	18	19	20
21	22	23	24	25	26	27	28	29	30
31	32	33	34	35	36	37	38	39	40
41	42	43	44	45	46	47	48	49	50

Write the missing number in each sequence.

1. 24 25 26 ◯ 28 29

2. 35 36 37 38 ☐ 40

3. ☐ 19 20 21 22 23

4. 41 ☐ 43 44 45 46

5. 28 29 30 31 32 ◯

6. 43 42 41 ☐ 39 38

7. 29 28 ☐ 26 25 24

8. 18 ☐ 16 15 14 13

9. 47 46 45 44 ◯ 42

10. ☐ 39 38 37 36 35

Colour in your score

36

Test 6 Subtraction: finding differences

Counting in jumps can help to find the difference.

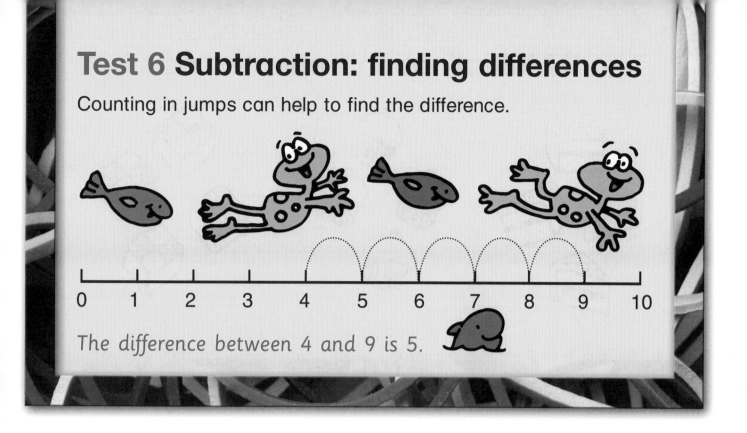

The difference between 4 and 9 is 5.

Write the differences between these pairs of numbers.

1. (4) (7) [ ]

2. (3) (9) [ ]

3. (5) (10) [ ]

4. (9) (2) [ ]

5. (8) (3) [ ]

6. (12) (7) [ ]

7. (9) (11) [ ]

8. (14) (8) [ ]

9. (13) (9) [ ]

10. (6) (11) [ ]

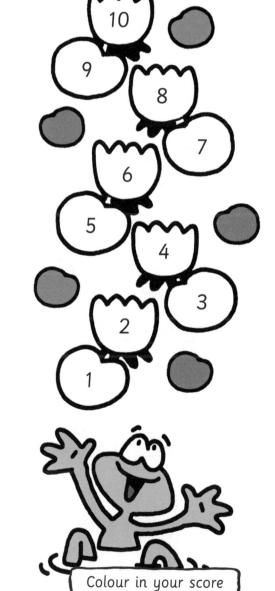

Colour in your score

Test 7 **Multiplication: repeated addition**

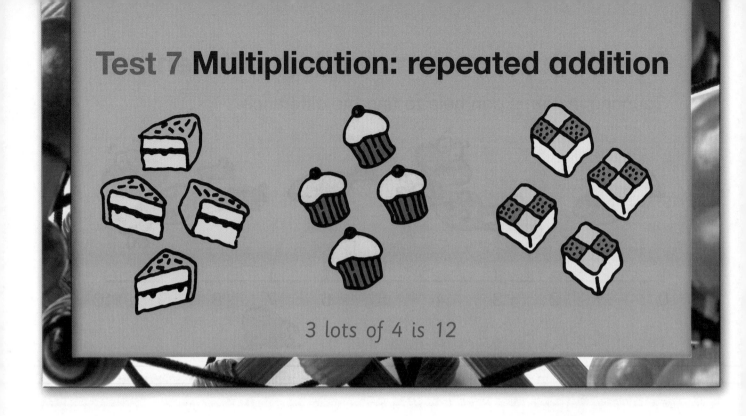

3 lots of 4 is 12

Write the answers.

1. 3 lots of 2 ⇨ ☐

2. 2 lots of 4 ⇨ ☐

3. 3 lots of 3 ⇨ ☐

4. 2 lots of 5 ⇨ ☐

5. 4 lots of 3 ⇨ ☐

6. 2 lots of 2 ⇨ ☐

7. 5 lots of 3 ⇨ ☐

8. 3 lots of 5 ⇨ ☐

9. 4 lots of 2 ⇨ ☐

10. 2 lots of 3 ⇨ ☐

Colour in your score

Test 8 Division: sharing

These sweets are shared equally.

15 sweets among 3 children ⟹ 5 each.

Write the answers.

1. 12 shared by 2 ⟹ [ ]

2. 8 shared by 4 ⟹ [ ]

3. 6 shared by 3 ⟹ [ ]

4. 10 shared by 2 ⟹ [ ]

5. 9 shared by 3 ⟹ [ ]

6. 12 shared by 3 ⟹ [ ]

7. 10 shared by 5 ⟹ [ ]

8. 6 shared by 2 ⟹ [ ]

9. 12 shared by 4 ⟹ [ ]

10. 8 shared by 2 ⟹ [ ]

Colour in your score

Test 9 Time (1)

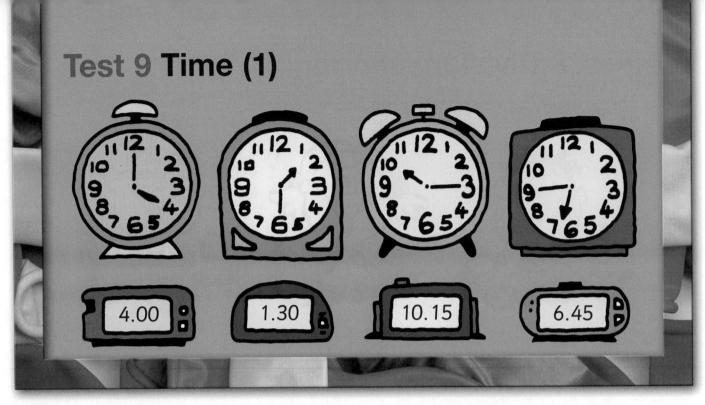

Write the times for each clock.
Choose from these times.

1.15	3.30	4.45	8.00	2.30
8.45	3.15	7.30	9.00	4.15

1.

2.

3.

4.

5.

6.

7.

8.

9.

10.

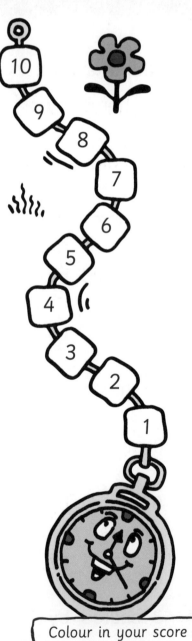

Colour in your score

Test 10 Data: block graphs

Colour this **graph** showing the favourite fruit of a group of children.

number of children

6
5
4
3
2
1

apples oranges pears grapes peaches bananas

fruit

How many children chose:

1. oranges? ☐

3. bananas? ☐

2. grapes? ☐

4. peaches? ☐

5. Which fruit was the children's favourite?

6. Which fruit was chosen by 3 children?

7. How many more children chose grapes than peaches? ☐

8. How many fewer children chose oranges than bananas? ☐

9. How many children chose pears and peaches altogether? ☐

10. How many children were there altogether? ☐

10
9
8
7
6
5
4
3
2
1

Colour in your score

Test 11 Breaking up numbers

$$30 + 7 = 37$$

$$20 + 4 = 24$$

Write the missing numbers.

1. 43 = ☐ + 3

2. 56 = 50 + ☐

3. 39 = 30 + ☐

4. 61 = ☐ + 1

5. 27 = ☐ + 7

6. 46 = 40 + ☐

7. 83 = ☐ + 3

8. 74 = ☐ + 4

9. 32 = 30 + ☐

10. 91 = ☐ + 1

Colour in your score

Test 12 Subtraction facts

This is a **function machine** for changing numbers.

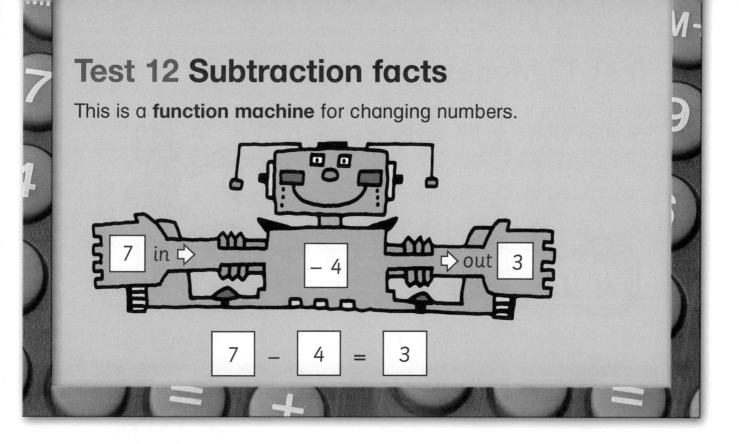

7 in ➡ −4 ➡ out 3

7 − 4 = 3

Write the missing numbers.

1. 8 − ☐ = 5

2. ☐ − 2 = 6

3. 7 − 3 = ☐

4. ☐ − 4 = 5

5. 9 − ☐ = 6

6. 8 − 2 = ☐

7. ☐ − 3 = 7

8. 8 − ☐ = 3

9. 6 − ☐ = 1

10. ☐ − 4 = 3

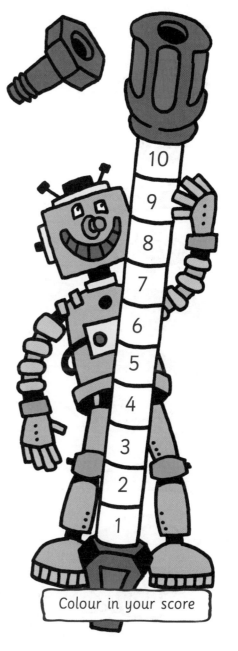

Colour in your score

43

Test 13 Money: giving change

When we work out **change** with **coins**, we often **start** with the **smallest value**.

65p

change: 35p

£1 is given for each toy. Write the change given.

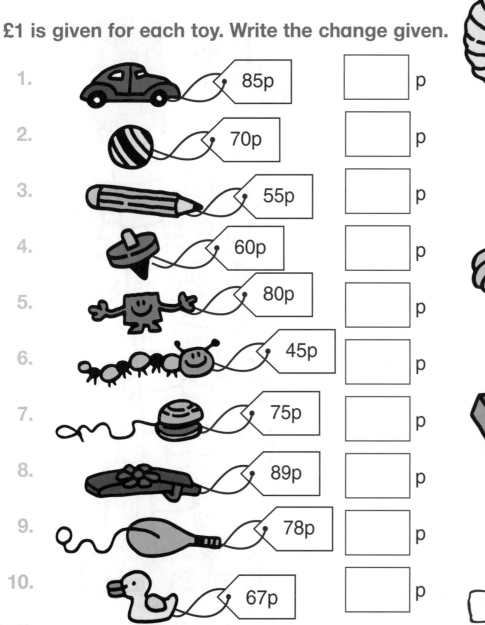

1. 85p ____ p
2. 70p ____ p
3. 55p ____ p
4. 60p ____ p
5. 80p ____ p
6. 45p ____ p
7. 75p ____ p
8. 89p ____ p
9. 78p ____ p
10. 67p ____ p

10
9
8
7
6
5
4
3
2
1

Colour in your score

Test 14 3-D shapes

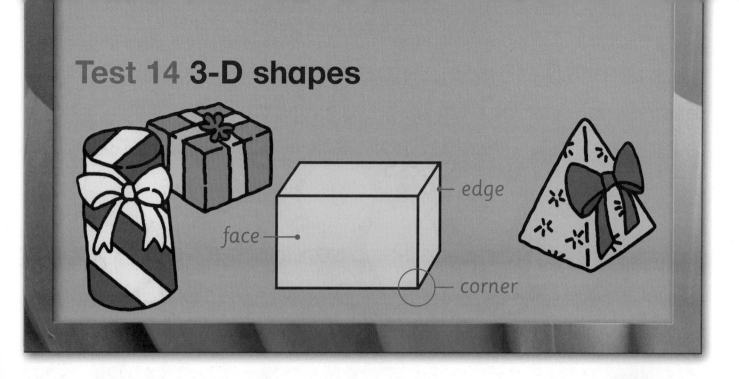

face — edge — corner

Name these shapes.

1.

_____

2.

_____

3.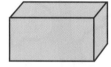

_____

4.

_____

5.

_____

6.

_____

How many faces have each of these shapes?

7. ☐ faces

8. ☐ faces

9. ☐ faces

10. ☐ faces

10
9
8
7
6
5
4
3
2
1

Colour in your score

45

Test 15 Counting patterns

Write the next number in the pattern.

1. 18 20 22 24 26 ◯

2. 34 36 38 40 42 ☐

3. 20 25 30 35 40 ◯

4. 15 18 21 24 27 ☐

5. 8 12 16 20 24 ◯

Write the missing number.

6. —14—☐—18—20—22—24—

7. —45—40—35—☐—25—20—

8. —9—12—☐—18—21—24—

9. —28—24—20—16—☐—8—

10. —☐—27—24—21—18—15—

Colour in your score

Test 16 Decade sums

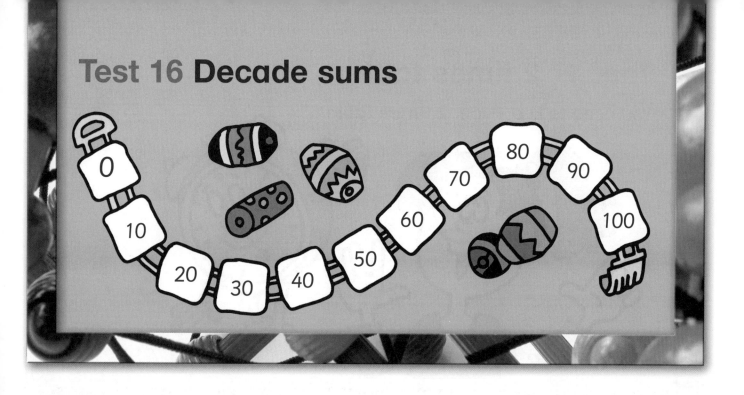

Answer these questions.

1. 40 + 20 = ☐

2. 30 + 30 = ☐

3. 30 + 10 = ☐

4. 40 + 60 = ☐

5. 50 + 40 = ☐

6. 20 + 20 = ☐

The three corner numbers add up to 100.
Write the missing number.

7.

10

100

30 ◯

8.

30

100

◯ 60

9.

◯

100

50 20

10.

◯

100

30 20

Colour in your score

47

Test 17 2 times table

You need to know your **2 times table**.

How quickly can you answer these?

1.   3 × 2 =

2.   7 × 2 =

3.   2 × 6 =

4.   8 × 2 =

5.   2 × 4 =

6.   2 × 9 =

7.   5 × 2 =

8.   1 × 2 =

9.   10 × 2 =

10.   2 × 2 =

Colour in your score

48

Test 18 Fractions: halves and quarters

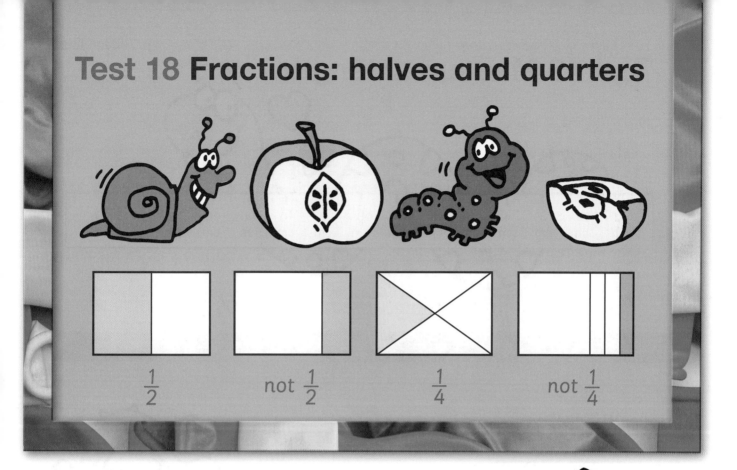

$\frac{1}{2}$ not $\frac{1}{2}$ $\frac{1}{4}$ not $\frac{1}{4}$

Colour $\frac{1}{2}$ of each shape.   **Colour $\frac{1}{4}$ of each shape.**

1. 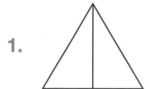   6.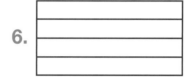

2. 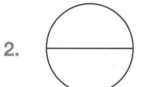   7.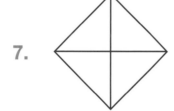

3.    8.

4.    9.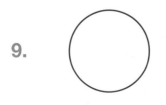

5. 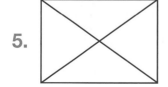   10.

Colour in your score

Test 19 Measures: length

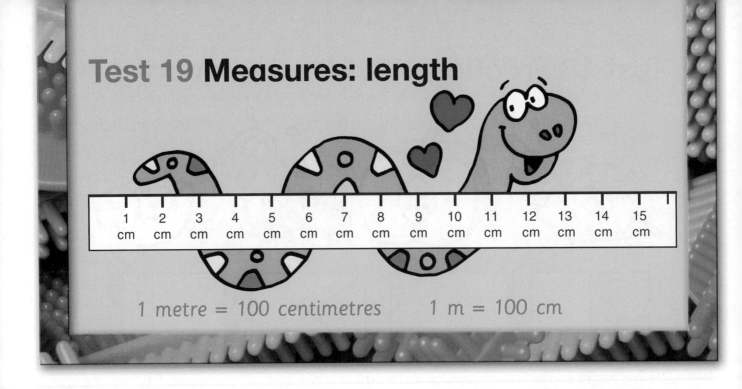

| 1 cm | 2 cm | 3 cm | 4 cm | 5 cm | 6 cm | 7 cm | 8 cm | 9 cm | 10 cm | 11 cm | 12 cm | 13 cm | 14 cm | 15 cm |

1 metre = 100 centimetres   1 m = 100 cm

Measure these lines.

1. _____ ☐ cm

2. _____ ☐ cm

3. _____ ☐ cm

4. _____ ☐ cm

5. _____ ☐ cm

Guess how long each worm is.

6. ☐ cm

7. ☐ cm

8. ☐ cm

9. ☐ cm

10. ☐ cm

10
9
8
7
6
5
4
3
2
1

Colour in your score

Test 20 Data: pictograms

This **pictogram** shows the pets owned by a group of children.

dogs	🐕 🐕 🐕 🐕 🐕
cats	🐈 🐈 🐈 🐈 🐈 🐈
rabbits	🐇 🐇 🐇
mice	🐁 🐁 🐁
fish	🐟 🐟 🐟 🐟

How many children have a pet:

1. dog

2. fish

3. mouse

4. rabbit

5. cat

6. How many more cats are there than fish ?

7. How many fewer rabbits are there than dogs?

8. How many mice and cats are there altogether?

9. How many fish and dogs are there altogether?

10. How many pets are there altogether?

Colour in your score

Test 21 Comparing and ordering numbers

Use this **number line** to help you **compare** and **order numbers**.

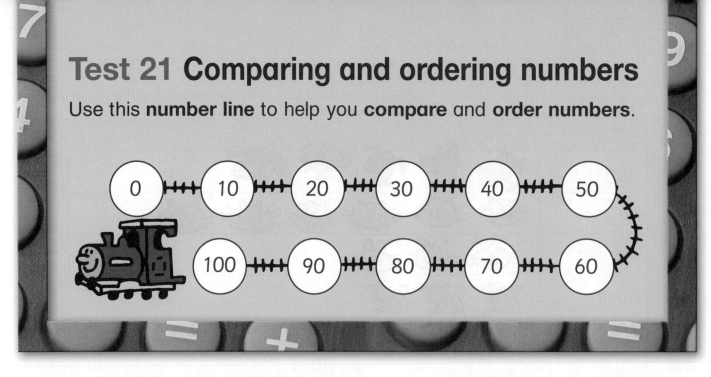

Circle the bigger number in each pair.

1. (46) (61)   2. (68) 83

3. (39) 41   4. 93 39

5. 57 54

Write the numbers in order starting with the smallest.

6. (18) (34) (27) (41) [   ]

7. (61) (52) (59) (62) [   ]

8. (38) (41) (52) (37) [   ]

9. (51) (53) (59) (54) [   ]

10. (72) (69) (64) (70) [   ]

Colour in your score

Test 22 Addition and subtraction

This **number trio** makes **addition** and **subtraction** facts.

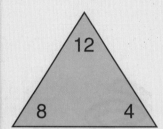

| 8 | + | 4 | = | 12 |

| 4 | + | 8 | = | 12 |

| 12 | − | 4 | = | 8 |

| 12 | − | 8 | = | 4 |

Write the addition and subtraction facts for each of these number trios.

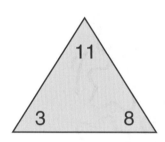

| 3 | + | 8 | = | 11 |

| 8 | + | 3 | = | 11 |

1. | 11 | − | ☐ | = | 8 |

2. | 11 | − | 8 | = | ☐ |

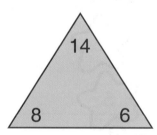

3. | 8 | + | ☐ | = | 14 |

4. | 6 | + | ☐ | = | ☐ |

5. | ☐ | − | 6 | = | ☐ |

6. | ☐ | − | 8 | = | ☐ |

7. | ☐ | + | 7 | = | ☐ |

8. | ☐ | + | 9 | = | ☐ |

9. | 16 | − | ☐ | = | ☐ |

10. | ☐ | − | ☐ | = | 7 |

Colour in your score

53

Test 23 Problems

To work out a **missing number** use the other numbers to help you.

$\square$ + 6 = 15

Something add 6 equals 15.

9 + 6 = 15

Write the missing numbers.

1. $\square$ + 3 = 11

2. 6 + $\square$ = 12

3. $\square$ + 7 = 15

4. $\square$ + 4 = 12

5. 9 + 6 = $\square$

6. 8 + $\square$ = 16

7. 4 + $\square$ = 11

8. $\square$ + 7 = 10

9. 5 + 8 = $\square$

10. 9 + $\square$ = 18

Colour in your score

10
9
8
7
6
5
4
3
2
1

54

Test 24 Shapes

These shapes have a line of **symmetry**.

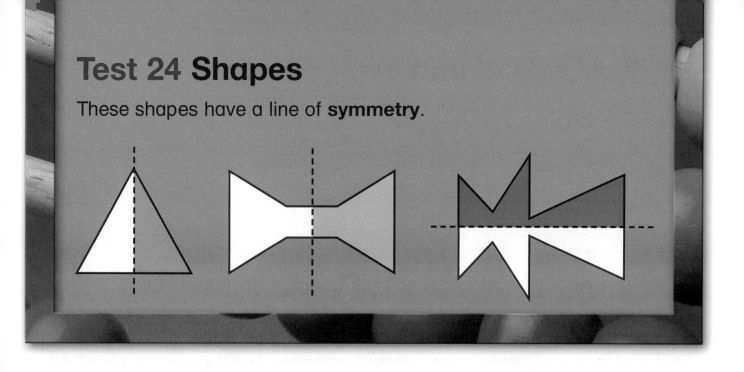

Draw the lines of symmetry on these shapes.

1.

2.

3.

4.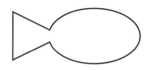

5.

Name these shapes.
Tick them if they are symmetrical.

6.
_____ ☐

7.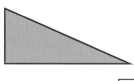
_____ ☐

8.
_____ ☐

9.
_____ ☐

10.
_____ ☐

Colour in your score

Test 25 Odd and even numbers

even	2	4	6	8	10	12	14	16
odd	1	3	5	7	9	11	13	15

Write the next even number.

1. 22

2. 28

3. 36

4. 44

5. 40

Write the next odd number.

6. 39

7. 41

8. 27

9. 35

10. 43

10
9
8
7
6
5
4
3
2
1

Colour in your score

Test 26 Money and place value

Do you know the shapes, values and sizes of these coins?

Write each of these totals.

1. 20p 20p 10p £1 ⇨ £ [.]

2. 50p 50p 20p 10p ⇨ £ [.]

3. 20p 20p 50p £1 ⇨ £ [.]

4. £1 £1 20p 10p ⇨ £ [.]

5. £1 50p 20p 10p ⇨ £ [.]

6. £1 £2 20p 10p ⇨ £ [.]

7. 10p 20p 20p £2 ⇨ £ [.]

8. £2 £2 50p 20p ⇨ £ [.]

9. 50p 50p 20p £2 ⇨ £ [.]

10. 20p 50p £1 £1 ⇨ £ [.]

Colour in your score

Test 27 **10 times table**

You need to know your **10 times table**.

How quickly can you answer these?

1. 4 × 10 =

2. 10 × 3 =

3. 7 × 10 =

4. 1 × 10 =

5. 5 × 10 =

6. 10 × 6 =

7. 10 × 2 =

8. 8 × 10 =

9. 10 × 10 =

10. 9 × 10 =

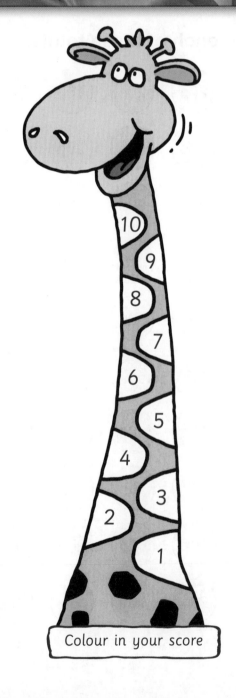

Colour in your score

Test 28 Fractions of quantities

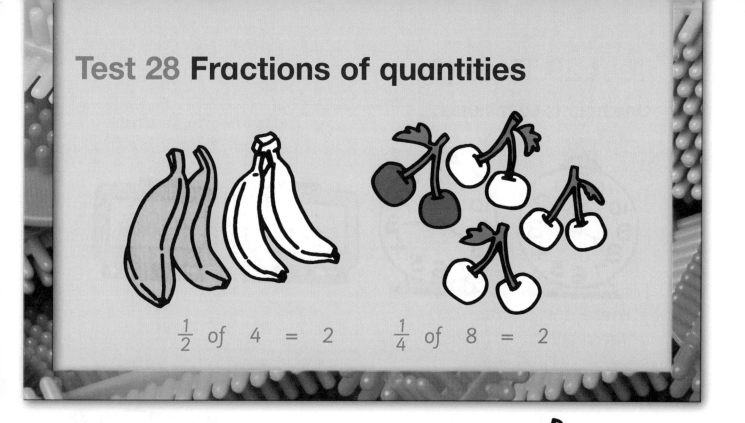

$$\frac{1}{2} \text{ of } 4 = 2 \qquad \frac{1}{4} \text{ of } 8 = 2$$

Find half of each of these.

1.  $\frac{1}{2}$ of 6 = ☐

2.  $\frac{1}{2}$ of 10 = ☐

3.  $\frac{1}{2}$ of 12 = ☐

Work out the answers.

4. $\frac{1}{2}$ of 8 = ☐

5. $\frac{1}{2}$ of 20 = ☐

6. $\frac{1}{2}$ of 14 = ☐

7. $\frac{1}{4}$ of 20 = ☐

8. $\frac{1}{4}$ of 12 = ☐

9. $\frac{1}{2}$ of 18 = ☐

10. $\frac{1}{4}$ of 4 = ☐

Colour in your score

Test 29 Time (2)

One hour is **60 minutes**.

There are 15 minutes between these times.

There are 15 minutes between these times.

Write how many minutes there are between each of these times.

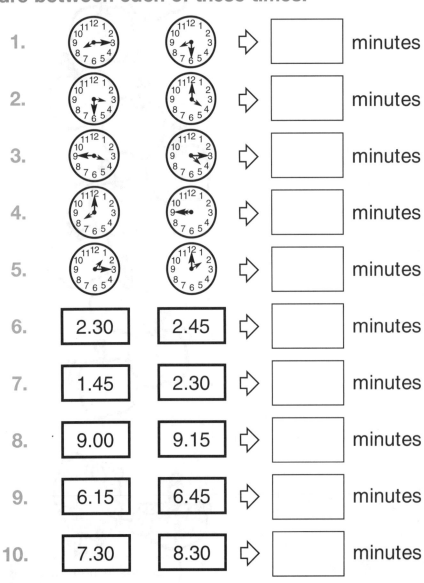

1. ⇨ [ ] minutes

2. ⇨ [ ] minutes

3. ⇨ [ ] minutes

4. ⇨ [ ] minutes

5. ⇨ [ ] minutes

6. | 2.30 | 2.45 | ⇨ [ ] minutes

7. | 1.45 | 2.30 | ⇨ [ ] minutes

8. | 9.00 | 9.15 | ⇨ [ ] minutes

9. | 6.15 | 6.45 | ⇨ [ ] minutes

10. | 7.30 | 8.30 | ⇨ [ ] minutes

Colour in your score

Test 30 Data: tables

This **table** shows the colours of children's tops.

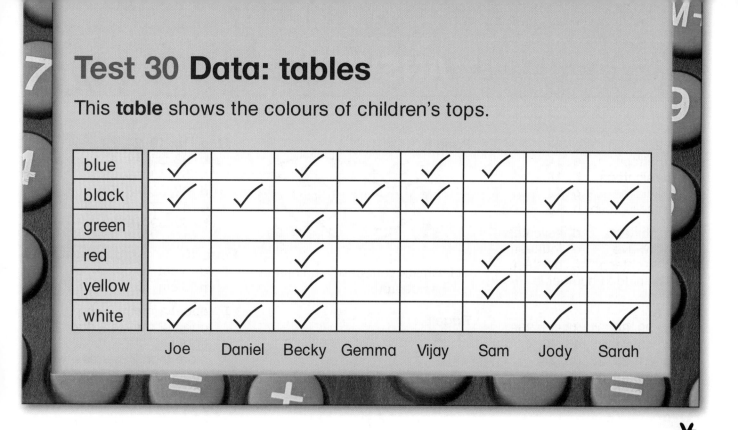

	Joe	Daniel	Becky	Gemma	Vijay	Sam	Jody	Sarah
blue	✓		✓		✓	✓		
black	✓	✓		✓	✓		✓	✓
green			✓					✓
red			✓			✓	✓	
yellow			✓			✓	✓	
white	✓	✓	✓				✓	✓

Look at the table and answer these questions.

1. Who has the most colours in their top? _____

2. Who has a black and white top? _____

3. What colours are in Sam's top? _____

4. Who has green in their top? _____

5. Who has no black in their top? _____

6. How many have white in their top? _____

7. How many have no blue in their top? _____

8. How many have 3 colours in their top? _____

9. Who has a single coloured top? _____

10. Who has no black or white in their top? _____

Colour in your score

ANSWERS

Page 2

I **a** fifteen **b** eighteen **c** eleven **d** seventeen **e** 14 **f** 19 **g** 12 **h** 16

II **a** thirteen **b** twelve **c** eighteen **d** seventeen **e** fourteen **f** nineteen

The hidden number is 11

Page 3

I
a 29, 30, 31, 33, 35, 36
b 40, 41, 42, 45, 46, 47
c 31, 30, 28, 25, 24
d 50, 49, 47, 46, 42, 41
e 17, 18, 19, 21, 23, 25

II a

	5				
14	15		17		
24	25	26	27	28	29
	35	36	37	38	
	45	46			

b

22	23		25	26
32	33	34	35	36
42		44	45	

c

6	7	8		10
		18	19	20
		28	29	30
	37	38	39	40
			49	50

Page 4

I **a** 9 **b** 12 **c** 12 **d** 9 **e** 12 **f** 14 **g** 13 **h** 15 **i** 11 **j** 13 **k** 13 **l** 14

II **a** 12 **b** 18 **c** 14 **d** 15 **e** 16 **f** 13 **g** 10 **h** 20 **i** 19 **j** 11

17 is the star coloured in.

Page 5

I

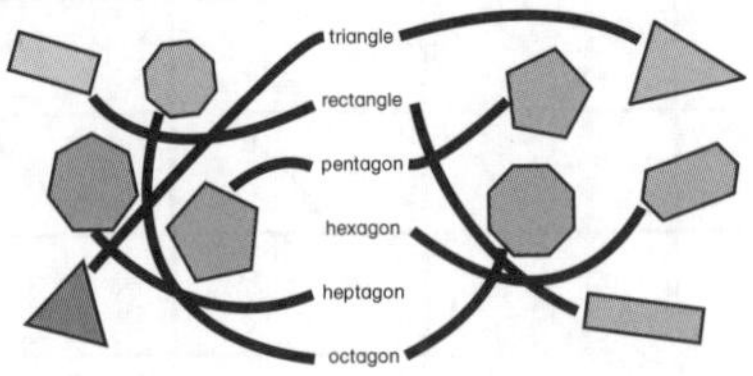

II Check your child's colouring is accurate.

Page 6

I **a** 7 **b** 8 **c** 5 **d** 9 **e** 12 **f** 11

II
a 8 – 2, 10 – 4, 12 – 6, 14 – 8, 9 – 3
b 14 – 7, 11 – 4, 9 – 2, 15 – 8, 13 – 6

Page 7

I **a** 38 **b** 54 **c** 79 **d** 62 **e** 87

II

T	W	E	N	T	Y	E	F
N	S	I	X	T	Y	F	I
I	Y	G	N	H	V	O	F
N	E	H	E	Y	I	R	T
E	S	T	H	I	R	T	Y
T	R	Y	M	L	F	Y	E
Y	S	E	V	E	N	T	Y

Page 8

I
a $5 + 8 = 13$ $8 + 5 = 13$ $13 - 5 = 8$ $13 - 8 = 5$
b $6 + 9 = 15$ $9 + 6 = 15$ $15 - 9 = 6$ $15 - 6 = 9$
c $9 + 8 = 17$ $8 + 9 = 17$ $17 - 8 = 9$ $17 - 9 = 8$

II $4 + 5 = 9$ $8 + 3 = 11$ $7 - 5 = 2$ $12 - 4 = 8$ $8 - 7 = 1$ $6 + 3 = 9$

Page 9

I

1	2	3	4	(5)	6	7	8	9	10
11	12	13	14	(15)	16	17	18	19	20
21	22	23	24	(25)	26	27	28	29	30
31	32	33	34	(35)	36	37	38	39	40
41	42	43	44	(45)	46	47	48	49	50
51	52	53	54	(55)	56	57	58	59	60
61	62	63	64	(65)	66	67	68	69	70
71	72	73	74	(75)	76	77	78	79	80
81	82	83	84	(85)	86	87	88	89	90
91	92	93	94	(95)	96	97	98	99	100

II
a 14, 19, 24, 29
b 32, 37, 42, 47
c 53, 58, 63, 68
d 28, 38, 48, 58,
e 47, 57, 67, 77
f 59, 69, 79, 89

Page 10

I **a** 5 cm **b** 6 cm **c** 8 cm **d** 2 cm **e** 10 cm **f** 11 cm

II about 1 metre – child
about 2 metres – door
about 10 cm – pencil
more than 2 metres – wall
about 50 cm – book

Page 11

I **a** 16 **b** 16 **c** 12 **d** 18 **e** 14 **f** 19 **g** 18 **h** 14 **i** 13

II
a There are many possible solutions, check your child's additions total 13.
b There are many possible solutions, check your child's additions total 18.

Page 12

I **a** 24 **b** 40 **c** 48 **d** 62 **e** 56 **f** 37 **g** 59 **h** 31 **i** 89 **j** 93

II

19	24	32	48	85	33	34	26	18	70	96	73	34	26	14
23	6	61	16	51	27	58	35	43	19	34	85	58	21	43
42	30	25	40	10	7	94	65	24	46	52	17	92	80	19
85	27	41	93	28	43	62	97	12	21	33	29	31	52	21
17	35	43	8	32	76	44	81	16	54	36	28	56	74	45

10 stars were collected.

Page 13

I

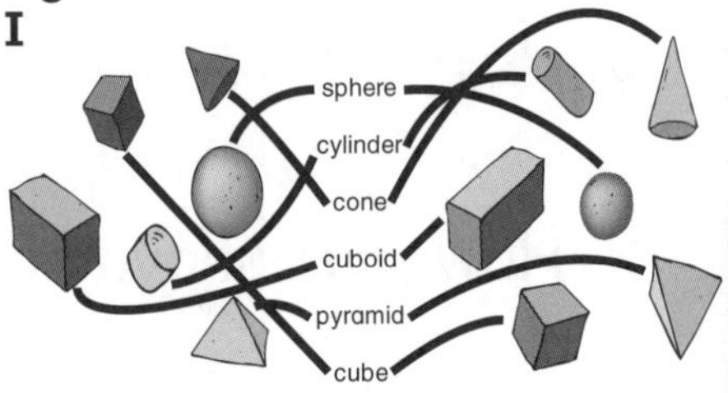

II
a 2 square faces, 4 rectangle faces
b 6 square faces
c 1 square face, 4 triangle faces
d 2 circle faces and a curved face

Page 14

I Less than 1 kg – biscuits, grapes, slipper, crisps
More than 1 kg – dog, potatoes, multi-pack of beans, encyclopedia

II a 4 kg   c 3 kg
   b 8 kg   d 9 kg

Page 15

I a 4   d 60   g 9
   b 50   e 3   h 70
   c 7   f 40   i 9

II

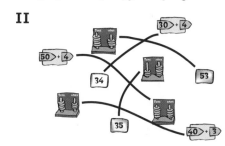

Page 16

I a 1.15   d 7.15   g 4.15
   b 6.45   e 12.45
   c 11.30   f 3.00

II a 30 minutes
   b 45 minutes
   c 15 minutes
   d 30 minutes

Page 17

I a 6, 6   d 8, 8
   b 10, 10   e 9, 9
   c 15, 15   f 16, 16

II a 12   b 18

Page 18

I

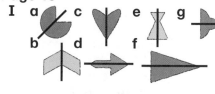

II

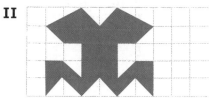

Page 19

I Answers can be any 2 numbers, in order, between:
   a 42 and 50   f 76 and 83
   b 61 and 73   g 94 and 98
   c 57 and 64   h 88 and 92
   d 82 and 91   i 57 and 61
   e 66 and 71

II a 39p, 42p, 58p, 61p, 85p
   b 37 kg, 39 kg, 69 kg, 73 kg, 76 kg
   c 23 cm, 28 cm, 32 cm, 38 cm, 80 cm
   d £29, £35, £53, £62, £65

Page 20

I a 4   b 3   c 4   d 5

II a 3   b 5   c 5   d 2

Page 21

I a Jo   d Jack
   b 5   e 4
   c Ali and Ryan

II Check your child's graph is accurate.

Page 22

I Spring: March, April, May
Summer: June, July, August
Autumn: September, October, November
Winter: December, January, February

II a 7   e 60   h 2
   b 12   f 60   i 4
   c 52   g 14   j 3
   d 24

Page 23

I $3 \times 2 \rightarrow 6$ $4 \times 2 \rightarrow 8$
   $5 \times 2 \rightarrow 10$ $6 \times 2 \rightarrow 12$
   $2 \times 2 \rightarrow 4$ $10 \times 2 \rightarrow 20$
   $8 \times 2 \rightarrow 16$ $7 \times 2 \rightarrow 14$
   $1 \times 2 \rightarrow 2$
   a $9 \times 2 \rightarrow 18$

II a 6, 14, 8, 2, 12
   b 20, 4, 16, 10, 18
   c 16, 12, 18, 14, 10

Page 24

I Check your child has halved each shape accurately and coloured one-half. There is more than one solution to some of the shapes.

II a 3   b 4   c 5   d 2

Page 25

I Less than 1 litre – glass of squash, cup of tea, medicine bottle, soup bowl
Greater than 1 litre – bottle of drink, fish tank, washing-up bowl, bucket

II a 7 litres   d 4 litres
   b 9 litres   e 2 litres
   c 6 litres   f 8 litres

Page 26

I a 20p, 20p, 2p
   b 50p, 2p, 1p
   c £1, 10p, 5p
   d 10p, 5p, 1p
   e 50p, 20p, 10p
   f £1, 20p, 20p

II a 15p   b 33p   c 41p

Page 27

I a 24, 26   d 36, 41
   b 32, 35   e 16, 12
   c 17, 15

II Ask your child what the difference is between the numbers in each sequence and check the numbers are correct.

Page 28

I a 15, 12, 40, 16, 20
   b 8, 70, 18, 25, 30
   c 4, 90, 30, 20, 35
   d 10, 80, 50, 6, 45

II $2 \times 1 = 2$, $2 \times 2 = 4$,
   $2 \times 3 = 6$ or $2 \times 4 = 8$
   $4 \times 5 = 20$
   $10 \times 8 = 80$, $20 \times 4 = 80$,
   $40 \times 2 = 80$ or $80 \times 1 = 80$
   $7 \times 5 = 35$ or $5 \times 7 = 35$
   $2 \times 9 = 18$, $9 \times 2 = 18$,
   $3 \times 6 = 18$ or $6 \times 3 = 18$
   $5 \times 8 = 40$ or $8 \times 5 = 40$

Page 29

I Check your child has divided each shape accurately into four parts and coloured one-quarter. There is more than one solution to some of the shapes.

II There are many solutions, check only 2 parts of each badge is coloured in.

Page 30

I a 5 pencils   d 20m
   b £1.50   e £1.10
   c 9 eggs

II a 5   c 8
   b 11   d 5

Page 31

I a £5   c £5   e £4
   b £8   d £5   f £6

II

Page 32
1. 34
2. 46
3. 28
4. 72
5. 89
6. twenty-three
7. fifty-six
8. ninety-one
9. sixty-seven
10. forty-nine

Page 33
1. 11
2. 14
3. 14
4. 16
5. 14
6. 17
7. 16
8. 12
9. 18
10. 19

Page 34
1. 62p
2. 19p
3. 33p
4. 82p
5. 18p
6. 73p
7. 54p
8. 28p
9. 86p
10. 39p

Page 35
1. 3
2. 6
3. 5
4. 8
5. 4
6. quadrilateral/ square
7. pentagon
8. triangle
9. hexagon
10. rectangle/ quadrilateral

Page 36
1. 27
2. 39
3. 18
4. 42
5. 33
6. 40
7. 27
8. 17

9. 43
10. 40

Page 37
1. 3
2. 6
3. 5
4. 7
5. 5
6. 5
7. 2
8. 6
9. 4
10. 5

Page 38
1. 6
2. 8
3. 9
4. 10
5. 12
6. 4
7. 15
8. 15
9. 8
10. 6

Page 39
1. 6
2. 2
3. 2
4. 5
5. 3
6. 4
7. 2
8. 3
9. 3
10. 4

Page 40
1. 9.00
2. 7.30
3. 4.15
4. 3.30
5. 8.00
6. 2.30
7. 8.45
8. 1.15
9. 4.45
10. 3.15

Page 41
1. 1
2. 5
3. 4
4. 2
5. grapes
6. apples
7. 3

8. 3
9. 4
10. 17

Page 42
1. 40
2. 6
3. 9
4. 60
5. 20
6. 6
7. 80
8. 70
9. 2
10. 90

Page 43
1. 3
2. 8
3. 4
4. 9
5. 3
6. 6
7. 10
8. 5
9. 5
10. 7

Page 44
1. 15p
2. 30p
3. 45p
4. 40p
5. 20p
6. 55p
7. 25p
8. 11p
9. 22p
10. 33p

Page 45
1. cylinder
2. cone
3. cuboid
4. sphere
5. pyramid
6. cube
7. 5
8. 5
9. 6
10. 6

Page 46
1. 28
2. 44
3. 45
4. 30
5. 28
6. 16

7. 30
8. 15
9. 12
10. 30

Page 47
1. 60
2. 60
3. 40
4. 100
5. 90
6. 40
7. 60
8. 10
9. 30
10. 50

Page 48
1. 6
2. 14
3. 12
4. 16
5. 8
6. 18
7. 10
8. 2
9. 20
10. 4

Page 49
1.
2.
3.
4.
5.
6.
7.
8.
9.
10.

Page 50
1. 4 cm
2. 2 cm
3. 6 cm
4. 7 cm
5. 8 cm
6. 4 cm
7. 5 cm
8. 7 cm
9. 8 cm
10. 6 cm

Page 51
1. 5
2. 4
3. 3
4. 3
5. 6
6. 2
7. 2
8. 9
9. 9
10. 21